OPERATION FINITO!

A Canadian Scientist came to Africa and settled in the rugged North Eastern Province in the Republic of Kenya. He set up a Laboratory which to the locals was a centre for the study of livestock diseases. The locals were very happy because an outbreak had completely wiped out all their livestock a few years before. He had promised to make all animal diseases a thing of the past .The government had given him a large tract of land where he would carry out his experiments. It was a secretive place which was well guarded and had a big wall around. The land was also surrounded by a canal which was full of crocodiles that made sure no unauthorized person gained entrance.

But Dr. Kevin Camp Daddy had other ideas. He had invented a way of manufacturing babies whose growth rate was one month for ten years and then normal growth set in after three months when the creature was thirty years old. He had bumped on to the idea when he was studying how he could save some endangered species from extinction. The success entered his head and he decided to bring forth a human race that was free from all the vices that afflict people today.

He had also made a chemical that could wipe out the human race in cities, towns and villages all over the world when released. He wanted to wipe out the all human beings from the face of the world and replace it with another creation of Super Kids. Where he is the father of the universe! The idea of him having a hand in all the children coming to this world motivated him even more. And Africa was a place where he would work without many nosy people trying to poke their noses in a business that was strictly his! Getting permits and the entire required infrastructure was also easier in Africa than any other part of the world. All you needed was to register a company and threw a few politicians as the Board of Directors and you were home and dry!

He had also formed an abduction squad to abduct a hundreds of women to be locked in his laboratory. Their ovulation was to be chemically regulated and then he would use his own sperms to produce his Super Kids. At the rate of one thousand per month! He had reckoned that he would be able to fill the whole world in less than ten years. The kids would marry and produce more super kids. The human race will be changed forever. And he will be the master of the bloody change!

But first things first! The first agenda was to capture the women that would provide the initial eggs. These were to be put in captivity and their ovulation regulated chemically to happen on a daily basis. It was not a hard thing for he had experimented with monkeys and the results were a resounding success. The abduction squad was to be complimented with lucrative job offers where advertisement will be placed all over the world. The ladies coming for interviews would also be abducted and kept in captivity.

The second part was to release the chemical into the world. The germs were transported through the air and he had projected that the world population would be wiped out within a week. It was only him, the captured women and his workers who would survive the initial onslaught. But he would take care of them immediately the second phase of the plan was over. The world will then be left to him and his super kids who will form a new generation of people. He will declare himself the King of the universe and everything would revolve around him from that time onwards.

As he sat looking at the future, he laughed out loudly. Nobody in this world had any idea of what lay in store for them. Nobody knew what Dr. Daddy was capable of doing. And the bastards will never know what hit them when he eventually rolled out his elaborate plan! He will laugh his head off watching thousands, then millions and finally billions perishing. Presidents leaving the whole world under his feet!

****

Three hundred well dressed women lined up for the interviews at the Royal International Hotel. The advertisement had said that there were openings for all cadres in a newly established hotel in the middle of Kenya's largest National Park. Some had been attracted by the huge salaries where others were just thrilled to work in the bushes with wild animals surrounding them at all times. All of them had similar characteristics. They were not only smart but very beautiful. Their intelligence was also above average. Dr. Daddy wanted a generation of clever people. His fantasy world did not have a place for foolish, ugly and dirty people!

The interviews were just routine questions and none could read in mischief in the insistence that one was supposed to be free from any family engagements to get the job. They were told that once employed, many contacts with the outside world were very much discouraged for it affected their level of production. After the long and grueling interviews, one hundred women from many different parts of the world were selected. They were immediately sent to the airport to be transported to their new work stations for orientation and the start of their duties. The last time that they were to see the outside world!

At the same time, the abduction squad had already captured over a hundred women. Dr. Daddy was pleased with the progress of his plan and was now ready to set things rolling. All the women were upon arrival drugged under the pretence of immunization against Malaria and locked to different cells where the treatment for regulated ovulation started immediately. Many responded quite well and those who were unlucky enough not to respond to the treatment were thrown into the canal where the crocodiles had a feast of their lives!

It took almost three months for Dr. Daddy to be satisfied that all was well. He had more than a thousand ovules ready for fertilization and production of the first bunch of the super kids. They were three very hectic months where he was forced to terminate the lives of seventeen women whose ovulation failed the test of the times and ten men who had found easy meat and raped several women as they guarded and fed the ladies. Preparations for the first thousand cages for the thousand super kids were the only thing standing between the start of the project.

Dr. Daddy would not take any chances. He was going to cage them and study them before releasing them to the bloody world. He did not want them turning into cannibals or any other social misfits, making life hard for everybody. But the construction was going on earnestly.

Things would start happening in less than two months.

****

The first Kid to come out had a very big head and one eye on the forehead. He looked like a ghost. The legs were thin and long and the body was not proportional. Dr. Daddy waited for three days and was surprised by the rapid growth and development of this ungainly creature. The growth rate was very much okay but now the shape was not what he had hoped for. He therefore decided to terminate the bloody thing immediately. The carcass... It was not a body and was thrown to the crocodiles which enjoyed another feast.

The second kid turned out to be exactly what the crazy doctor was looking for. He was talking within three days. The growth rate was so fast that even the workers who lived in the Lab were terrified. Dr. Daddy had not prepared them for this and so many wanted to leave immediately. They were conveniently done away with and their bodies used to supplement the diet of the crocodiles.

The first thousand kids were now put into life and Dr. Daddy had a rough time trying to feed the many new mouths. He had budgeted for their food using normal rations. But these were not normal kids. They were eating twice as much as normal human beings. He had to release them immediately to go and fed for themselves and this meant that he didn't have time to study them better. He also had to clear the world sooner than he had planned to create room for the new owners of the universe.

A few kilometers from the Laboratory was a Refugee camp holding thousands of people from a neighboring country! The place where Dr. Daddy had decided to do his first experiment! With very devastating results! He had mixed the dangerous chemical with Maize Flour. He then donated packets of the same to some women who lived in the camp and waited for the results of his dirty work.

It did not take long for that evening; reports of mysterious deaths in the refugee camp had filled the whole country. Soon the whole world was reporting the news of the disaster which was proving more than a handful. Thousands of bodies lay dead. Their eyes bulging out through some unseen pressure! Nobody had the faintest idea as to what was happening. Apart from Dr Daddy! He smiled as he watched the news in the television. The newsmen who rushed to cover the unfolding ammargedon were also dropping dead. This was not only serious but catastrophic.

It took only two days to wipe out a whole camp with tens of thousands of refugees. And it was spreading to other neighboring towns and villages. Nobody had been found alive to describe what they felt after an attack by the strange disease. The only common denominator was the eyes which seemed to pop out of their sockets. The scientists who had been sent by the American government to check into the outbreak also died before they could even get near the place.

Dr. Daddy could not believe the success of his maiden attack. He launched his second attack by sending missiles to Nairobi the capital city of Kenya, Kampala in Uganda and then Dar el Salaam in Tanzania. Two hours later, BBC was announcing the devastating events unfolding in East Africa. People were dropping everywhere and dying like locusts. Operation Finito was on and nothing was going to stop it from happening. No media house was risking sending its people to the East African Countries. They were now relying on Satellite pictures. Things were now very grim. They situation was quickly getting out of hand.

The whole of East Africa was now under attack. And the chemical was spreading like wild fire. It would take three days for the chemical to become impotent. Three days that was taking people down in their thousands. The presidents of the three countries had been evacuated to South Africa. Corpses now littered everywhere in their millions.

Dr. Daddy launched several other missiles to Rwanda, Burudi and Congo. The East and Central Africa was now at the mercy of the doctor. Three days later, ten more missiles found their way to North Africa, South Africa and the West Africa. Over one hundred million people had already died. Twenty nine African heads of State had met their death while the rest had been evacuated to Europe. Dr Daddy followed all these through his Satellite Television and always laughed loudly when more countries kept going down.

A week of relentless attacks had left Africa without any living soul. The only people that survived were those that had travelled to other parts outside Africa when hell broke loose. It was time to release the first bunch of the super kids to the world. They were divided into bunches of two hundred each and were now ready to be transported to the four corners of the continent and Central Africa. Exactly ten days after the first missiles were launched, five airplanes set off to deliver their human cargo to capitals' around Africa. They were to be left to fed on their own and start a new generation. Dr Daddy had ensured the women were more than men in the new set up!

The second bunch of super kids was in their third day of incubation when their "kinsmen" were released to this bloody world. Dr. Daddy had planned to wait for ten days before he launched another attack. But that evening, BBC and CNN showed footages of the planes dropping the kids. This meant that American Intelligence would try to follow up and see what was happening. Immediate action was needed to avoid any attack of the program from Europe and America. They had to be hit before they launched their attack.

That night, missiles laden with the now
infamous chemical were launched headed
to Europe, America and Australia. The
effect was immediate. Both CNN and BBC
stopped operating immediately. It would
only mean that the chemical had done its
work. More were directed to Asia, India and
South America. The whole world was now
under attack. Almost a billion people had
died and more were meeting their creator
each and every passing day. All animals
living in the air and dry land were also
dead. The only animals that had survived
were those that live in water.

The president of the United States was
sitting in the Situation Room with top
intelligence people trying to figure out
what was happening. They were following
the unfolding end of the world when all the
machines seemed to "hang'. They had no
idea what was happening. Suddenly, he
saw his Chief General of Staff convulse and
then his eyes literally popping out of their
sockets. He only had time to think of his

wife who was asleep when he was overcome by a sudden force that seemed to constrict his head. He tried to shout but nothing came from his mouth. And he fell down! His eyes dancing out of their sockets! Very dead!

Meanwhile, the second bunch was ready for dispatch. Dr. Daddy immediately set them off to Europe and America. Others were sent to Australia. He was going to leave the Middle East without people for the purpose of settling the best of the best so that they could benefit from the oil. The third bunch was in incubation and the project was going on well. The only issue now was to round off all his workers after the second delivery and finish them off so that only the new generation of human beings would remain in this world.

****
****

Thousands of Miles in Mars, the Green people who inhabited the place were watching the unfolding scenario on earth with a lot of interest. They had always known that human beings were going to ruin their earth one day. But the number of deaths was something that they had never anticipated even in their wildest dreams. They had witnessed many calamities and wars. But nothing had prepared them for the mass destruction of human life that was going on.

Many a times, people from Mars had been accused of many bad things happening in earth. They had been accused of shooting down airplanes and even coming down to earth and raping human beings. It was rumored that they had even fathered many children some of which had turned out to be giants. Like Goliath! But the truth of the matter was that they had never interfered with the earth people. But what was happening now was too much for them.

It was also funny that all this was being done by a single human being whereas the other disasters were a combination of many things and war. Gurunero their leader called for a special meeting to discuss the issue and see how they could benefit from the whole thing. It was decided that an army of a thousand Green men be sent to earth and save it from further destruction. Their first assignment was to travel to Kenya and stop the manufacture of more super kids and then move around the globe hunting and killing all those who had been dispatched to various corners.

* *
* * * * * * * * * * * * * * *

In the meantime, Dr. Daddy was burning the mid night oil trying to figure out whether there had been any survivors in any part of the world. The fact that his chemical did not work in water was giving him more nightmares than ever. He knew that there might have been other people at sea during his attacks and this made him work extra hard to study satellite pictures from the entire world. After the second dispatch, he had summoned all his workers together for a briefing. Before he appeared in their meeting room, a pungent smell had filled the room and all had died within seconds.

The super kids had started marrying and getting children. The women were carrying their pregnancies for only seven days and the new race was multiplying again. At a very fast rate! The good thing was that the chemical was killing and then accelerating the decomposition of the bodies at a very fast rate. Were it not for the sparse population, a visitor could not have realized that something was amiss by a casual glance.

****
****

The third bunch was released from the Lab and was left just ten kilometers away. They were very excited to come from their cages and set off in different directions. Two by two! That night, Gurunero and his troops touched ground in their plate-shaped flying Saucers. Dr. Daddy realized that something was amiss the moment, he returned to his work station. His satellites had caught sight of the flying saucers and they disturbed him a lot. He sat by the switch board and tried to make sense out of everything.

Suddenly, the screen flickered twice and then went black. He punched the back-up and what he saw made him jump involuntary. His kids were being attacked and killed by some Green men who had mouths with long snouts. Like those of a pig! He quickly mounted his missiles and launched them towards the invaders. His super kids started falling down one by one. But the bloody invaders just jumped on and continued with their unholy mission. The third bunch was gone before he could even think of anything much to help them.

Then the Green Army turned and headed towards the Lab. He had to do something. But what could he do? Here was a bunch of people who seemed to have supernatural powers. People who walked through a chemical that had wiped out the entire universe! And they were now coming for him! To destroy him and his work! To kill him! If only he had developed some form of communication with his super kids. He could have summoned them to come and assist him. Without any help, he hoped that his crocodiles would do something. He looked at the direction they were coming from and an idea came.

He rushed to the cells and removed three women. He shot them and put their bodies in a cravat that led to the canal surrounding the lab. He cut them until they had bled enough and then used some water to push them down to the canal. Their blood would attract the crocodiles to that general area and the Green men would face the music if they tried to cross over. Or so he thought! He was in for a rude surprise. The Green Men literally walked over water. The crocodiles pulled the corpses further away and gave way to the new comers who walked as if they owned the universe. Dr. Daddy moved deeper in to the Lab and waited. They were not going to get away with it just then. He locked himself in a Gas chamber and released the most concentrated chemical to his compound. The women died there and then. But the Green men trudged on unperturbed!

Their control point in Mars informed Gurunero that there was only one human being alive in the entire building. They also knew that it would be extremely difficult to reach the bastard. Therefore, a missile was to be launched from the command base in Mars that was going to destroy the whole structure. The Green Army was therefore commanded to start their retreat immediately. Dr. Daddy tried very hard to release more concentrated chemical but it ended up to naught. He was shortly surprised to see the funny creatures' retreating and he thought that the chemical had overpowered them. As he sat up to celebrate, the whole building shook with an impact that sent all stones crumbling down in smithereens. Dr. Daddy was buried in holocaust. With a diabolic snarl plastered in his bloody face!

The Green Army then started jumping up and down singing victory songs in a language them alone could understand. Their first assignment was over and that called for celebration. Anybody watching them dancing with their long tails and snouted mouths would have been forgiven for thinking that it was a Devil's Dance. It continued for a very long time until they were exhausted. They had carried their own food from home and they ate as they pecked each other with their snouts. It was a sight to behold!

Gurunero then led his troops back to their flying saucers for their journey back home. They had been told to return immediately the mission was over so that they could study the spread of the super kids and plan the next assault which will leave the earth with no inhabitants paving way for the Green people from Mars coming down and spreading their Kingdom. They had seen man landing in the moon and also almost coming to their land. It was now their turn to live beyond their original home. It was their time to colonize the world and enjoy living in earth. They had lived respecting human beings and had never interfered with their lives. But the human beings had become too clever for their own good. Through sheer madness, they had wiped themselves out of the face of the earth!

The partying in Mars continued from several days. They then had a meeting after which top Seer Bonoko blessed all the soldiers and gave them the verdict of their gods. The mission to earth was going to be successful. The spread of the super kids was also studied and the Green Army organized themselves accordingly. Ten thousand soldiers with their hunting dogs were set off to go and complete the mission that was going to bring planet earth to the hands of the Green people! They were to hunt down all the super kids and kill them. Their hunting dogs were trained to hunt and follow even the faintest scent. They could scent the fart of a mosquito!

****
***

The assault was launched immediately. Tens of thousands of flying saucers set off from Mars and were soon headed to different corners of the world. The super kids had not spread very much. Their hunting was therefore scheduled to take less than a month. Gurunero had instructed his people to take the least time possible and ensure that nobody was left alive. A thousand hunting dogs would ensure that no place was left out as they flushed out the super kids.

The first attackers started with Kenya, the cradle of the super kids. They set on their mission with great zeal and soon were attacking the super families. And killing them in droves! In a span of three days, they had wiped out all the kids that had been dispatched to Kenya. They had done it faster than it had been anticipated and therefore had to wait for two more days before a signal for their next mission was to come. It was then that a calamity struck!

In their excitement to come and capture planet earth, they had forgotten to carry provisions for their hunting dogs. The dogs had become very hungry and with nothing to eat, they started feasting on the corpses. They found it a sweet delicacy and feasted on the dead throughout the day. With nothing much to do, the attackers made merry and danced throughout which left them extremely tired and exhausted. Nobody noticed when the hunting dogs started changing!
The metamorphosis was dramatic. After feasting on the corpses, the dogs started growing big. The growth mechanism that had been set in super kids had been transferred to the dogs. They were now the size of the elephants. The attackers were

perplexed by the new development. Contact with their command base was one day away. They had no way of letting their comrades know about the new development. They had a situation in their hands which needed urgent attention.

What they could not know then was that the same scenario was unfolding everywhere. The hunting dogs were growing at a very alarming rate. There was confusion everywhere. The situation had to be arrested soon or it would go out of hand. Gurunero was worried that the dogs would soon get out of control.

Luckily, Seer Bonoko had seen what was happening and requested the Command Base to contact Gurunero. The gods had a second opinion of the expedition colonize planet earth. The unforeseen change of events was going to bring more havoc to them than good. They were to start going home before the dogs went hungry again and started feasting on them. Dr. Daddy was still pulling the strings. From hell!

In the meantime, hell had broken loose in Europe. The poor weather had made communication difficult and therefore the evacuation order did not reach them. The corpses had gone and the dogs had started looking for food. And the attacks on the Green men started earnestly! Their hunting dogs had started feasting on their masters. This was too much for them. They had never been prey to anything and this new development shocked them to the core. The hunter had now become the hunted!

The macabre sight of giant dogs feasting on green people with green blood shocked even the hardest souls in Mars. They had never seen something like this. The colonization that had anticipated was turning into their worst nightmare. And the orgy of killings continued. Some would fight back and kill some of the dogs.....which was immediately set upon by the other dogs for feasting!

Eerie barks and cries lent the air everywhere. There was crying and gnashing of teeth. No place was safe for any living creature in the world. And the giant dogs continued to reproduce and fill all corners of the world. Luckily for them, water animals had escaped Dr. Daddy's autocracies. They could hunt in want and continue living. Giant dogs roaming the plains, valleys and the huge buildings left behind by the now extinct human population.

In Mars, the Green people watched the scenario with dismay. What they had hoped would be their second home was now inhabited by their own hunting dogs. Dogs that had been transformed into giants. Giants that feasted on anybody and anything. Giants that have been created by the genius of a man who was long dead.

And in the end there was life.....And the life was giant dogs...And the dogs ruled the world. After the humans had wiped their race out of the face of the earth...And the dogs had dethroned the kingdom of their green masters! Amen.

THE END.

www.ingramcontent.com/pod-product-compliance
Lightning Source LLC
Chambersburg PA
CBHW070759180526
45168CB00004B/1676

REFERENCE

Mathematics of Finance (2015) Toye Adelaja.

NOTE:

Personal budget is not limited to students alone. It can be used by every individual irrespective of age.

Smith's Monthly Budget

Sources of Income	$
Work	500
Parents	200
Scholarship	250
Available funds/money	950

Monthly Expenses	
Necessity	
Phone bill	80
Rent	200
Car Insurance	40
Groceries	280
Gas	100
Wants	
Cloths	200
Latest movie	50
Tourism	100
Total monthly expenses	1050
Difference	**-100**

The available fund minus total monthly expense is -$100. This is a budget deficit. Smith has to make some adjustments to his budget in order to arrive at budget balance.

He should first consider reducing its expenses on wants and if the reduction in expenses on wants cannot give him budget balance he should go further by reducing variable needs expenditure. It is advisable for him to reduce expenses on clothes by $100 or expenses on latest movies and tourism by $50 each.

Phone bill	80
Rent	200
Car Insurance	40
Groceries	250
Gas	80
Wants	
Cloths	60
Latest movie	25
Tourism	30
Total monthly expenses	<u>765</u>
Difference	**285**

The available fund minus total monthly expense is $285. This is a budget surplus.

Where you prepare your budget and you arrive at a budget deficit, you need to make some adjustments to the expenses in order to arrive at a balance figure. This will be discussed in Step 5.

Step 5: Make Adjustment if needed

Where your budget is deficit, you need to make some adjustments to the expenses in order to arrive at budget balance.

You can adjust your expenses by first cutting down expenses you incurred on wants. If after reducing the expenses on wants, you are still unable to arrive at budget balance, you can go further by reducing your variable needs expenditure in the short- term and your fixed expenditures in the long-term.

Take a look at the example below:

If you have a monthly saving goal and you include the savings as part of your spending (cash out flow), it will be easier for you to meet up with the goal if you have added it to your budget.

It will be easier to prioritize your expenses if you classified your expenditure into 3 categories as mentioned above.

Expenses will be arranged according to the order of importance and necessity. By doing this, the least important expenses can be removed in order to balance your budge (step 5).

Step 4: Make Summation and Deduction

Compare your income and expenses by deducting the expenses from your income. Does the difference between your income and expenses give you a surplus or deficit. If you get surplus as the result, you are on the right side and you can invest in your future but if your result is deficit, read step 5 below.

My Monthly Budget

Sources of Income	$
Work	500
Parents	300
Scholarship	250
Available funds/money	1050

Monthly Expenses
Necessity

If you want your goals to be achievable, you must have reliable source or sources of income. Is your money coming from work? Is your money coming from investment? Is your money coming from student loan? Is your money coming from your parents? Is your money coming from scholarship? In a nutshell, you must have reliable sources and amount of income.

Step 3: Where is your money going?
You need to know where your money is going before you can prepare a budget. Check your bank account to know how you have been spending your money. Check how much you spent out of the cash you are holding. If you want to track the accurate records of your spending, ensure to record your expenditure on a daily basis.

Use a spread sheet to track and categorize your expenses for a month.
It is necessary to classify your expenses into 3 categories:

Fixed expenses: These include expenses such as rent, phone bill, etc that are fixed for each month. You must definitely incur these expenses.

Variable expenses: These expenses vary and not stable. They include gas, food, fuel, cloths etc. They are necessities.

Wants: These are non-essential expenses such as chips for refreshment, latest movie etc.

$$1.0398$$

A   = $57.70

You need to set aside $57.70 per month in order to meet up with the cost of the car.

Where:

FV is the future value of the ordinary annuity.

A is the equal amount to be paying at the end of each period.

r is the rate of compound interest.

n is the number of years of the payment or receipt.

PV is the present value of the ordinary annuity.

NOTE:

Take a look at the above goals. You will discover that it complies with SMART goal:

S = Specific

M= Measurable

A = Achievable

R = Relevant

T = Timely

Step 2: Where is my money coming from?

Short-term goal: Less than a year

Mid-term goal : one to three years

Long-term goal: More than five years

For example, assume that you want to buy a car immediately you graduate from school and you have 36 months to spend in school. You have to start saving for each month now. If the cost of the car is going to be $3,000 in three years, how much do you need to save per month in order to meet up with the cost? Rate of interest on savings is assumed to be 2% per month. Annuity or Sinking fund can be used to solve this problem.

Solution:

This is an ordinary annuity.

The amount you need to set aside at the end of each month is as calculated below:

$$FV = \frac{A\,[(1+r)^n - 1)]}{r}$$

$$\$3,000 = \frac{A\,[(1+0.02)^{36} - 1)\,]}{0.02}$$

$$= \frac{A[2.0398 - 1]}{0.02}$$

$$\frac{\$3,000}{1} = \frac{1.0398A}{0.02}$$

$$1.0398A = 3,000 \times 0.02$$

$$A = \underline{60}$$

CHAPTER FOUR

PERSONAL BUDGETING

What is personal budgeting?

Personal Budget is a financial plan that allocates future personal income to future personal expenses, savings, and debt repayment. It can also be defined as a future estimate of revenue, costs and resources. It can be monthly, quarterly and yearly.

How do I create Budget?

The following steps are involved in creating personal budget.

Step 1: What are my goals?

The first step in setting up a budget is to identify and set up your goals. You goal may be to pay a debt, to buy a house, to prepare for retirement age, to minimize the debt you graduate with, to save for car or your family. Budget may involve difficult choices but having a specific goal will simplify it.

Every financial goal you set should be a SMART goal:

S = Specific

M= Measurable

A = Achievable

R = Relevant

T = Timely

Your goals can be classified into three categories:

Private Pension Plans are the pension schemes that are being managed by corporation order than government. This is not reliable because corporations are being liquidated. Once corporation goes into bankruptcy, all your savings for building your future pension scheme go down.

How Much Would I Need?

How much you plan depends on your taste, your desired standard of living, your expenses including any medical costs and your target age of retirement.

How much you need to plan can also depend on the common old age at which people die in your community or environment. You can work toward this by preparing a sinking fund. The sinking fund has already been explained in chapter three of this book.

CHAPTER THREE

RETIREMENT PLANNING

Retirement is inevitable as death is inevitable to every human being. You cannot work through out your life span. There will be a time in which your energy and strength will be dwindled. At this stage of life, you will not be able to expend energy as you used to. You have no option than to retire from work. The period in which you will not be able to work is referred to as period retirement.

Even, if all of us are willing to retire comfortably, the complexity involved in retirement planning can be discouraging and intimidating. However, it only requires little assignment, savings and proper investments. Once you can do these you will rejoice at the end.

Why Retirement Planning?

We have to know the reasons for retirement planning before we discuss on other processes involved in retirement planning. The reasons for retirement planning will stimulate us to ensure that we absolutely plan for retirement. The following are the reasons for retirement planning:

1) Uncertainty of social security and Pension Benefits

Government is not finding it easy these days to implement social security and pension benefits. The numbers of employees that are contributing to the scheme are reducing and hence, the amount available to administer the pension is reducing. You need to plan for your own retirement if you don't want to be a victim of degrading in the scheme.

2) Private Pension plan

Save to Add Value to Your Money

When you save $1 in a bank account today, you will be having an amount higher than $1 tomorrow. This is what is called time value of money. The interest will be added to your savings in the bank. Your money in the bank will be increased by the addition of the interest. If you cultivate the habit of savings, you will add value to your money.

$$\$50{,}000 = \frac{A\,[(1+0.2)^{5}-1)]}{0.2}$$

$$\$50{,}000 \times 0.2 = A(1.2)^{5} - 1$$

$$\$10{,}000 \quad = A(2.4883 - 1)$$

$$\$10{,}000 \quad = A \times 1.4883$$

$$\frac{\$10{,}000}{1.4883} \quad = A$$

$$\$6{,}719 = A$$

The amount that will be kept aside annually is $6,719

Sinking Fund Schedule:

Years	a Balance b/f	b Interest $	c Sinking Fund $	a+b+c Balance c/d $
1			6,719.00	6,719.00
2	6,719.00	1,343.80	6,719.00	14,781.80
3	14,781.80	2,956.36	6,719.00	24,457.16
4	24,457.16	4,891.43	6,719.00	36,067.59
5	36,067.59	7,213.52	6,719.00	50,000.11

The sinking fund for each year is $6,719

There are some benefits of cash purchasing such as cash discount and other allowances. If you purchase by cash you stand a better chance of gaining than to purchase on credit. There is what is called cash discount. It means that if you purchase goods by cash, the cost of the goods will be reduced. This means that you have to pay less for the goods. For example, if the cost of the goods is $300 and a discount allowed of 5% is given for cash purchases. The cash discount will be 5% ×$300 = $15. You will eventually pay $285 instead of $300 for the goods. If you can save your money in order to purchase by cash, you will gain this cash discount in many circumstances.

Save for Sinking Fund

A sinking fund is a stream of equal savings you made for the purpose of acquiring property or fixed assets in the future. The assets can be houses, cars and other luxuries. You have to set aside the cash meant for the purchase of these assets.

Application of Annuity to Personal Decision

The following can be used as an example to explain sinking fund.

ILLUSTRATION 1

Mr. Joe needs to provide $50,000 to replace his machine in 5 years time, in order to provide this amount he decides to set aside equal amount annually, out of his salary. This amount is kept in savings account that yield 20% interest per annum. Find this amount, and the sinking fund schedule.

SOLUTION

The calculation of each amount that will be kept aside annually to meet this need is:

$$FV = \frac{A\left[(1+r)^n - 1\right]}{r}$$

CHAPTER TWO

SAVINGS

WHY DO YOU NEED TO SAVE MONEY?

What is your belief for saving money? Some people accept that they do not need to spare cash since they have enough to spend before the next earnings/salaries are paid. You may be asking yourself why is it necessary to save money since you have access to credit facility. In order to prove all these perceptions wrong, I sat down and put into written why you need to save money. The following are 5 important reasons you need to save money:

Save for Emergency Funds

You need to save for emergency funds. Some urgent financial needs such as repair of your car, payment for hospital bills and payment for all other unforeseen circumstances may occur. If you don't have required amount of money in your bank accounts to take care of all these situations, you may suffer unnecessarily. You need a continuous savings in order to build up emergency funds in addition to other insurance schemes you might have engaged in. It is never too late to start setting aside a percentage of your earnings or salary every month for the emergency funds. You can be saving 20% of your earnings or salary every month for these funds. If your monthly income is $800, your emergency fund will be 20% $\times$ $800 = $160, It means that you have to save $160 each month.

Save for Old Age

You need to save for the time you will not be able to expend energy to generate income. When you become old, you will be weak and you will not be able to work well as you used to do when you were younger. Save now and save your future as well as that of your family.

Save to Gain the Benefit of Cash Purchasing

Money owed you	xx	
Cash balance	xx	
Bank balance	xx	
Credit card balance(+)	<u>xx</u>	
	<u>xx</u>	
Total Assets	xx	xx
Liabilities		
cash borrowed	xx	
Accrual	xx	
Credit card balance(-)	<u>xx</u>	
	xx	
Car loans	xx	
Students loans	xx	
Mortgage balances	<u>xx</u>	
Less Total Liabilities	xx	<u>xx</u>
Net worth		xx

- Credit card balance (positive figure)
- Personal property (the resale value of your jewelry, etc)

Liability is the money you owed people.

The following are various categories of liabilities:

- Cash borrowed
- Services you enjoyed but not yet paid for(accrual)
- Car loans
- Students loans
- Credit card balance (negative figure)
- Remaining mortgage balances

The difference between your assets and your liabilities is your **net worth**. You begin to increase your net worth by reducing your liability and increasing your assets. You must endeavor to calculate your net worth often in order to know what you worth.

Here is an example of a format for the preparation of statement of financial position.

Statement of personal financial position as at 31st December, 2014.

Assets	$	$	$
Automobile value		xx	
Home value		xx	
Investments		xx	
Personal property		xx	
		xx	

The expenditure will be recorded on the debit side while income will be recorded on the credit side.

NOTE:

The recording of income and expenditure is different from the recording of cash book because, it involves recording of both cash transaction and non- cash transaction. This is called accrual basis in accounting.

1.2.2.3 Statement of Financial Position

Your personal net worth is the best way to determine what you worth and where you are in finance. Your net worth is calculated by deducting your liability (what you owe) from your assets (what you own). Statement of personal financial position shows all the assets you have and the liability you incurred as at a particular period. The components of statement of financial position are assets and liabilities.

Assets are those properties you have which cannot be consumed within a period less than a year. The following are various categories of assets:

- Automobile value (the resale value of your car)
- Home value (the resale value of your home)
- Your computers (the resale value of your computers)
- The amount of money owed you
- Your investment
- Your cash balance and bank balance

The cash and bank balance brought forward to the following month (February,2015) are $1,450 and $3,260 respectively.

Benjin's cash book for the month of February, 2015

Date	Particulars	Debit Side		Credit Side	
		Cash	Bank	Cash	Bank
		$	$	$	$
1	Balance b/f	1,450	3,260		
5	Friend		700		
5	Electricity bill				300
5	Water bill				100
6	Deposit		1,300	1,300	
	Balance c/d			150	4,860
	Total	1,450	5,260	1,450	5,260
	Balance b/d			150	4,860

NOTE:

The cash that Benjin is having in hand as at February 6, 2015 is $150 and the bank balance at the same date is $4,860.

1.2.2.2. Income and Expenditure Account

This account shows how you earn your income and how you utilize it. It has two opposite sides called debit side and credit side. It is used to determine the amount of surplus or deficit you have at the end of each month. If the total amount of income you realized at the end of the month is higher than the total amount you spent, you will have surplus but if the total of your expenditure at the end of the month is higher than the amount you realized as income at the end of the month, then you will have deficit.

He withdrew $1,200 in his bank account on January 30, 2015.

He paid for electricity bill of $300 in February 5, 2015 by check.

He paid for water bill costing $100 in February 5, 2015 by check.

He incurred $140 on consumption of fuel in January 26, 2015. He paid by check.

He deposited $1,300 to his bank account on February 6, 2015.

Benjin's cash book can be prepared from the above information.

Benjin's cash book for the month of January, 2015

Date	Particulars	Debit Side		Credit Side	
		Cash	Bank	Cash	Bank
		$	$	$	$
24	Salary		5,000		
24	Transport allowance	250			
26	Fuel				140
28	Dividend		100		
29	Food stuff				500
30	Cash withdrawn	1,200			1,200
	Balance c/d			1,450	3,260
	Total	1,450	5,100	1,450	5,100
	Balance b/d			1,450	3,260

NOTE:

It can be clearly seen that $1,200 withdrawn from bank was credited to bank column and debited to cash column.

One column cash book contains only cash transaction on both debit and credit sides while two columns cash book contains both cash and bank transaction on both debit and credit sides. The common type out of the two is two column cash book because many individual keep bank account.

The end result of preparing the cash book is to ascertain the cash and bank balances at the end of each month.

Note:

All transactions that occur in each month should be posted to the respective month in which they occur. For example all transactions that occur in January, 2015 should be posted to cash book for the month of January, 2015.

Example 1

Mr. Benjin is an engineer. He works with a construction company. The following are his transactions for the month of January and February, 2015.

Income:

His salary of $5,000 was paid into his bank account on January 24, 2015.

He received transport allowance $ 250 by cash on January 24, 2015.

A dividend of $100 was deposited into his bank account on January 28, 2015.

A friend that borrowed money from him paid into his bank accounts on February 5, 2015. The amount was $700.

Expenditure/out flow of cash

He bought food stuff worth $500 on January 29, 2015, by check.

Some people always forgot money in the pockets of their cloths and washed the money along with the cloths. This is a waste of money.

Some forgot money in the pockets of their cloths and gave the cloths out for laundry. This is a waste of money.

The solution to the above disadvantages is the keeping of records of personal finance. If records are kept frequently, you will know where each amount of your money is.

1.2.2. Types of Record Keeping

The following are the some necessary records that should be kept by every individual.

1) Cash Book

2) Income and expenditure Accounts

3) Statement of Financial Position

1.2.2.1   Cash Book

Cash book is a book that is used to record cash received, cash spent, cash lodged in a bank account and cash withdrawn from the bank. You can also use your cash book to record the movement of your cash as a business entity does.

Cash book contains two sides; both debit and credit sides. Debit side is used to record cash that comes in while credit side is used to record cash that goes out.

The cash book can be classified into two. They are:
- one column cash book
- two column cash book.

Record keeping is a pivotal element of personal finance. Record keeping is the methods of recoding every transaction involved in personal finance. It ranges from recording of salary or any other income received, recording of savings and recording of any money spent. Record keeping could be cumbersome to some people and it can be simple to some people. It depends on the way you perceive and practice it. Some people record every cent of their transaction while some ignore immaterial transaction.

It is advisable to record every transaction irrespective of the amount involved.

1.2.1. Reasons for Record Keeping

The following are the reasons for record keeping:

1) **To know where your income is coming from and to track it**

   Recording of your income enables you to know the source or sources of your income. It enables you to know whether you have received all the income you are entitled to receive from various sources.

2) **To know how much you have spent and what you spent on**

Some people spend money but unable to be accountable for the spending. They may not be able to explain how they spend there money. Some people may spend some amount of money and receive a balance from a vendor. They may not know where they keep the balance. If records are kept for the spending, you will be able to know whether you are having some balances somewhere or not. Some of the disadvantages of not keeping record of personal finance are as follows:

CHAPTER ONE

INTRODUCTION TO PERSONAL FINANCE

1.1.   Personal Finance

Personal finance is the whole processes involved in earning income, receiving gifts, saving money and spending money. Personal finance is very vital to every human's existence. It is like a blood flowing through the veins of human beings, and hence it should be handled with care and prominence it deserves.

As business finance is important to every business so is personal finance is important to every human. If a business is well financed, the business will be flourishing. If a person is well financed, the person will be well and have a bright future. Every business transaction is recorded on daily basis. The same thing should be applied to personal transaction. Personal transaction such as receiving of income, savings and spending of money should be recorded on daily basis.

Business entity borrows money to execute projects. The entity manages the loan and ensures that the loan is paid with interest as at when due. Human beings can also obtain loan and ensure that the loan is paid with interest as at when due.

1.2. Record Keeping

TABLE OF CONTENTS

Chapters	Contents	Pages
	Introduction	3
	Table of Contents	4
1	Introduction to Personal Finance	5
1.1	Personal Finance	
1.2	Record Keeping	
2	Savings	14
3	Retirement Planning	18
4	Personal Budgeting	20

INTRODUCTION

Personal finance is the whole processes involved in earning
income, receiving gifts, saving money and spending money.
Personal finance is very vital to every human's existence. It is
like a blood flowing through the veins of human beings, and
hence it should be handled with care and prominence it deserves.

ISBN (978 1517030810)

PERSONAL FINANCE AND MONEY

Toye Adelaja